AVALANCHES
AND LANDSLIDES

© Aladdin Books Ltd 1992

*First published in
the United States in 1992 by*
Gloucester Press Inc
95 Madison Ave
New York, NY10016

Design: David West
          Children's Book
          Design
Designer: Stephen Woosnam-
          Savage
Editor: Fiona Robertson
Researcher: Emma Krikler
Illustrator: Mike Saunders

Library of Congress Cataloging-in-Publication Data

Walker, Jane (Jane Alison)
     Avalanches and Landslides / by Jane Walker
        p.   cm. — (Natural disasters)
     Includes index.
     Summary: Studies the causes, effects, and
consequences of avalanches and landslides.
Includes historical facts and patterns, as well as
an examination of humankind's role in causing
these disasters.
     ISBN 0-531-17363-1
     1. Avalanches—Juvenile literature. 2.
Landslides—Juvenile literature. 3. Man—Influence
on nature—Juvenile literature. [1. Avalanches. 2.
Landslides. 3. Man—Influence on nature.]
     I. Title. II. Series
     QC929.A8W35   1992
     551.57'848—dc20   92-9085   CIP AC

Printed in Belgium

Natural Disasters

AVALANCHES
AND LANDSLIDES

JANE WALKER

GLOUCESTER PRESS
New York · London · Toronto · Sydney

CONTENTS

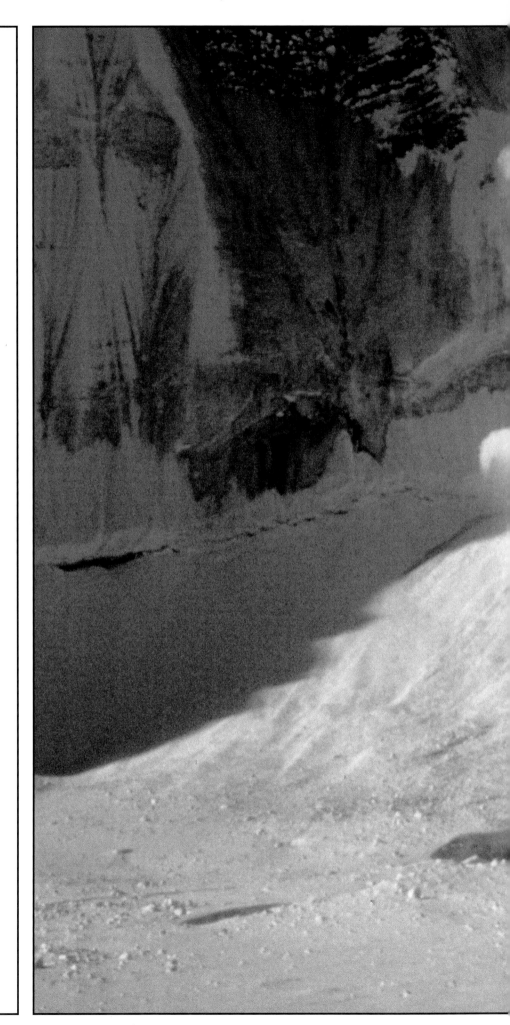

INTRODUCTION

A huge mass of snow, rock, or mud crashes down from a mountain or hillside at high speed. It can engulf and destroy entire towns and villages in minutes. With a thunderous roar, avalanches and landslides strike quickly and often without warning, and their impact can be catastrophic.

Hundreds of avalanches occur each day on remote snowy mountaintops throughout the world. However, their effect is only really felt in the populated mountain regions. Here, the whirling mass of ice and snow will crush to death or bury alive any person or animal unlucky enough to be caught in its path.

The introduction of improved warning systems can reduce the effects of avalanches and landslides. But we must also try to control the human activities that can trigger these natural disasters.

Although the term "avalanche" can mean the fall of any material – snow, soil, rocks, ice, or volcanic lava – it generally refers to a falling mass of snow. Dry-snow avalanches can travel at speeds of up to 225 mph. The melting snow of a wet-snow avalanche moves more slowly, at about 20 mph.

Wet-snow avalanche

Wet-snow avalanches often begin in rocky areas. Here, the snow melts around rocks that have warmed up in the sunshine. The fresh, wet snow can set as hard as concrete when the avalanche finally comes to a halt.

Melting snow exposes rocky surface

New snow layer

Dry-snow avalanche

The force of a dry-snow avalanche can equal that of a hurricane. As the powdery snow hurtles downhill, a destructive blast of air may be sent out ahead of the avalanche.

Air blast

Direction of powder

6

New snow layer

Direction of slab

Slab avalanche
As a slab avalanche hurls itself
down a slope, the front of the
slab starts to break up. Slab
avalanches usually occur on
slopes that are protected from
the wind, where the snow
collects in deep piles.

Snow rolls into
balls

WHAT IS AN AVALANCHE?

An avalanche is a huge mass of ice
and snow which breaks away from
the side of a mountain and surges
downward at great speed. The
greatest avalanches probably occur
on the high peaks of the Himalayas.
However, those which cause the
highest death toll fall in the
populated valleys of the Alps.

Scientists have grouped avalanches
into three main kinds: wet-snow
avalanche, dry-snow avalanche, and
slab avalanche. Wet-snow avalanches
usually occur in the spring, when the
loose, melting snow forms into large
boulders of snow as it rolls downhill.
More deadly are the dry-snow
avalanches, which either slide along
close to the ground, or lift off the
ground completely and swirl through
the air, often hundreds of feet high. In
a slab avalanche, a huge chunk of
solid, sticky snow breaks away from a
slope. It slides across a layer of loose
snow crystals lying beneath the
surface.

AN AVALANCHE BEGINS

Most avalanches begin either during or soon after a snowstorm. As each new layer of snow settles on the ground, it binds itself to the existing layers that are anchored to the mountainside. The additional weight of a heavy snowfall can prevent the snow from gripping onto the layers of snow underneath, and trigger an avalanche. Avalanches also occur during the spring, when melting snow seeps down through the surface. This creates a slippery layer on which the snow can slide.

The steepness of the slope can also affect the speed of the avalanche. Steep, rocky slopes help to anchor the snow. The smooth surface of gentler grass-covered slopes allows the avalanche to reach great speeds.

Heavy snowfalls
The weight from deep piles of falling or drifting snow can produce an avalanche.

Explosions
An avalanche can be set in motion by an explosion at a mine or quarry, by the noise of a low-flying plane, or even by a clap of thunder.

Avalanches are not only caused by natural factors, such as the amount and quality of the snow, or the sudden arrival of strong winds. Once conditions are suitable, an avalanche can be started by the slightest movement, such as a falling icicle or twig, or by a small animal crossing a slope.

Cornices
Hard winds drive the snow and pack it together to form a cornice, which hangs over the very top of the slope. When a cornice falls, it can trigger an avalanche.

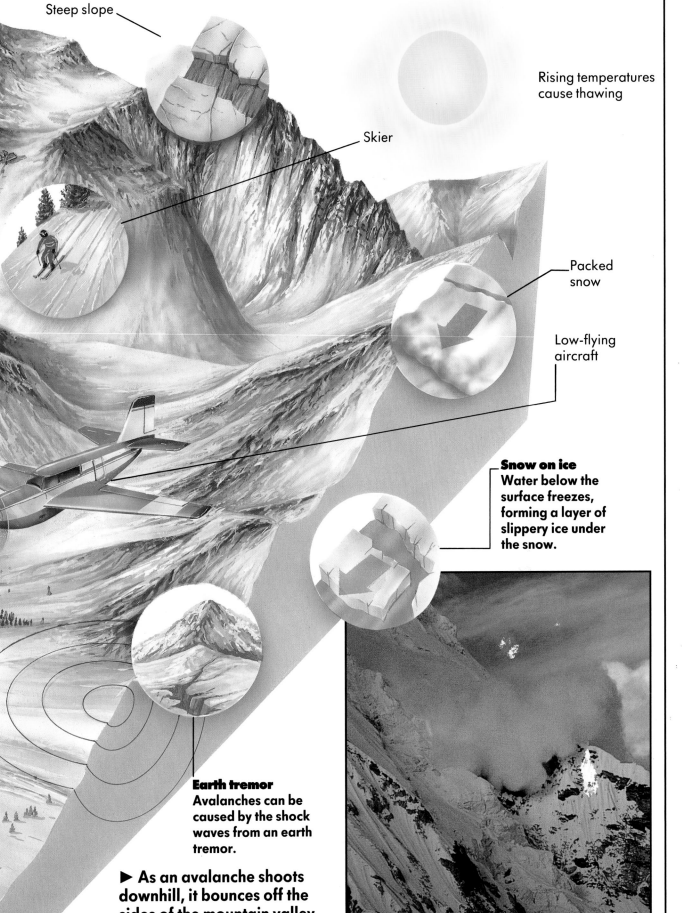

Steep slope

Skier

Rising temperatures
cause thawing

Packed
snow

Low-flying
aircraft

Snow on ice
Water below the
surface freezes,
forming a layer of
slippery ice under
the snow.

Earth tremor
Avalanches can be
caused by the shock
waves from an earth
tremor.

▶ As an avalanche shoots
downhill, it bounces off the
sides of the mountain valley,
creating clouds of icy snow
dust (right).

THE WHITE DEATH

When an avalanche strikes, more than 130 million cubic yards of snow and ice blast down a mountainside. As the ice mass falls, it collects large amounts of debris such as rocks and tree stumps on its way. In a populated mountain area, whole villages are crushed, hundreds of people and animals are buried alive, power and water supplies are cut off, and roads and railroad lines disappear – all in a matter of minutes.

The world's greatest single avalanche disaster occurred in 1962 in Peru. More than 3,500 people died and eight villages and towns were destroyed in just 7 minutes. When the avalanche finally came to rest, after a journey of almost 10 miles, the pile of snow and ice was over 60 feet deep.

Trapped under the hard, packed snow, an avalanche victim can barely move and is unlikely to survive for more than a couple of hours. Victims die from the cold, from a lack of oxygen, or from the injuries that occurred when they were first struck. Only about 5 percent of all avalanche victims are rescued alive.

6:18 p.m.
The deadly torrent of ice and rocks misses Yungay, but crushes the village of Ranrahirca, killing 2,700 people.

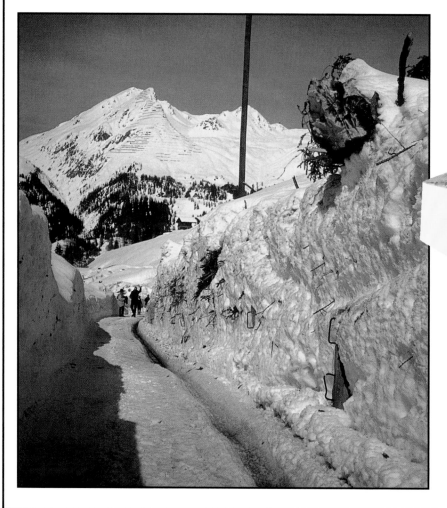

◄ In the mountain village shown left, a path has been cut through the wall of snow and debris caused by a recent avalanche.

Melting snow plunges down from the north-facing summit of the 22,205 feet high Nevado Huascarán mountain.

6:13 p.m. The falling snow loosens millions of tons of ice from Glacier 511.

6:15 p.m. The avalanche flattens Yanamachico and nearby villages, leaving 800 dead and only 8 survivors.

The 1962 avalanche disaster in Peru occurred when a huge chunk of ice and snow broke off the country's tallest mountain, Nevado Huascarán. The avalanche roared downhill at speeds of up to 60 mph, demolishing nearby villages with a deluge of mud and ice over 80 ft thick.

THE ALPS

Each year, tens of thousands of avalanches occur in the European Alps, the world's most densely populated mountain region. The inhabitants of the numerous Alpine towns and villages live under the constant threat of avalanches, which is made worse by certain natural conditions that prevail there.

The warm dry Föhn wind which blows through the deep Alpine valleys raises temperatures suddenly, causing the snows to melt rapidly. The smooth surface of the Alpine meadows lubricates the path of the avalanches, which travel at terrifying speeds. In 1898, an avalanche at Glärnisch in Switzerland raced downhill at around 250 mph.

During the 1950-51 "Winter of Terror" in Switzerland, over 1,100 avalanches killed a total of 98 people.

Albertville

Moutiers

Mont Blanc

Chamonix

FRANCE

Les Arcs

Valmorel

The Alps stretch for about 745 miles through seven countries, including France, Italy, and Switzerland. The area of the French and Italian Alps shown above is one of the most popular skiing regions.

Les Menuires

Méribel

Courchevel

12

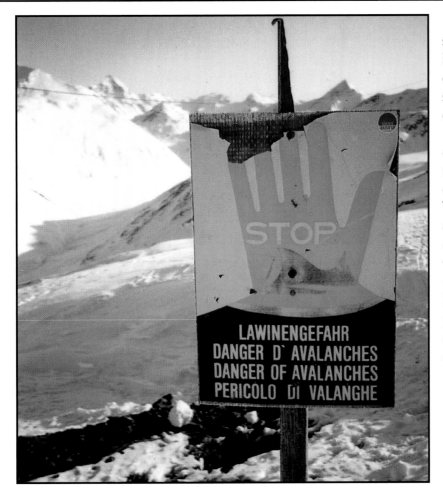

◀ Avalanche warning signs, such as the one shown left, inform skiers if there is a danger of avalanches. The ski trails, called pistes, that are at risk are then closed.

Experienced skiers who choose to ski "off piste" on unmarked trails run the risk of setting off an avalanche. The weight of a single skier on a slope can open up a crack in the hard crust on the snow's surface. This releases a huge slab which breaks away, often at a point above the skier, offering little chance of escape.

ITALY

Tignes

Val d'Isère

▶ More and more people visit the Alps each year. New hotels, ski lifts, electricity lines, and roads are being built to meet the increased demand for facilities. Avalanches in these developed areas are now more likely to cause widespread damage and many deaths.

WHAT ARE LANDSLIDES?

A landslide is the name given to the downhill movement of large amounts of soil, mud, rocks, and other debris. There are three different kinds of landslide: falls, slides, and flows.

In a rockfall, huge boulders crash down a steep slope, often breaking up into smaller pieces on reaching the ground. When rockslides occur, the chunks of rocky material slide over the ground as quickly as flowing water. Debris slides, which consist of thin layers of loose soil and smaller rocks, travel in the same way.

Mudflows and earthflows are wet landslides. A mass of mud and water flows down from the upper slopes, picking up debris lying in its path.

Curved sliding surface

Slump
A slump occurs when heavy rock and soil on a steep slope or cliff collapse in a series of curving movements. The blocks of slumped material tilt back toward the slope.

Blocks tilt backward

Rockslide
Rockslides involve the downward movement of rock debris or large blocks of rock. They often occur after heavy rains and in areas where large numbers of trees have been removed from the slopes.

Large blocks become detached from slope.

Mudflow streams out of bowl-shaped area

Mudflow

Mudflows are the most liquid kind of landslide. They occur in mountainous regions, desert areas, and on steep-sided volcanoes. They occur after heavy rains and can travel at speeds of up to 55 mph.

Rocks break off along lines of weakness

Large tongue of mud, water, and fine debris

Rockfall

A rockfall occurs when rocks descend at high speed, often falling freely through the air, down a steep slope or cliff face. The rocks may have been loosened by frost or ice. On reaching the lower slopes, a rockfall may break up into a moving torrent of debris.

Steep, bare slope

Loose debris on slope

Flooding

Flooding is a common aftereffect of landslides. The debris from a landslide can fill a riverbed. The river may then burst its banks, causing flooding (right).

In Utah, in 1982-83 heavy rains caused a huge mudslide that damned Spanish Folk Canyon. A lake of water was created, which drowned the nearby town of Thistle.

LANDSLIDE CAUSES

A landslide begins when the loose material on the surface of a slope becomes unstable. Water is one of the key factors in triggering landslides. After heavy rainfall or melting snow, the surface soil becomes saturated. Water seeps through the top layer, making the layers underneath very slippery. The water increases the weight of the surface material – soil, small rocks, and sediment – and weakens the forces that bind it.

As the soil loses its grip on the slope, it can no longer resist the downward pull of gravity. On steep slopes, the water-filled mass forms into a slide. On gentler slopes, it becomes a flow of earth and debris.

Erosion, where the lower layers of a slope are cut or worn away, is another major cause of landslides. Erosion can result from natural factors, such as the action of the sea or rivers, or from artificial ones, such as mining and excavating. A landslide itself can be a major force of erosion, especially on steep slopes where the tree cover has been removed.

Landslides are also associated with earthquakes and volcanic activity. After the devastating earthquake in 1964 in the state of Alaska much of the destruction resulted from landslides.

▼ **In the Trisuli Valley in Nepal (below), the soil has been eroded from the steep mountainsides. This erosion is partly due to tree-felling operations in the area.**

The upper slopes of this valley have slipped downward in a landslide. The river that flowed through the valley wore away the lower slopes, leaving the upper ones without support.

Water-soaked layers of shale help to lubricate the landslide.

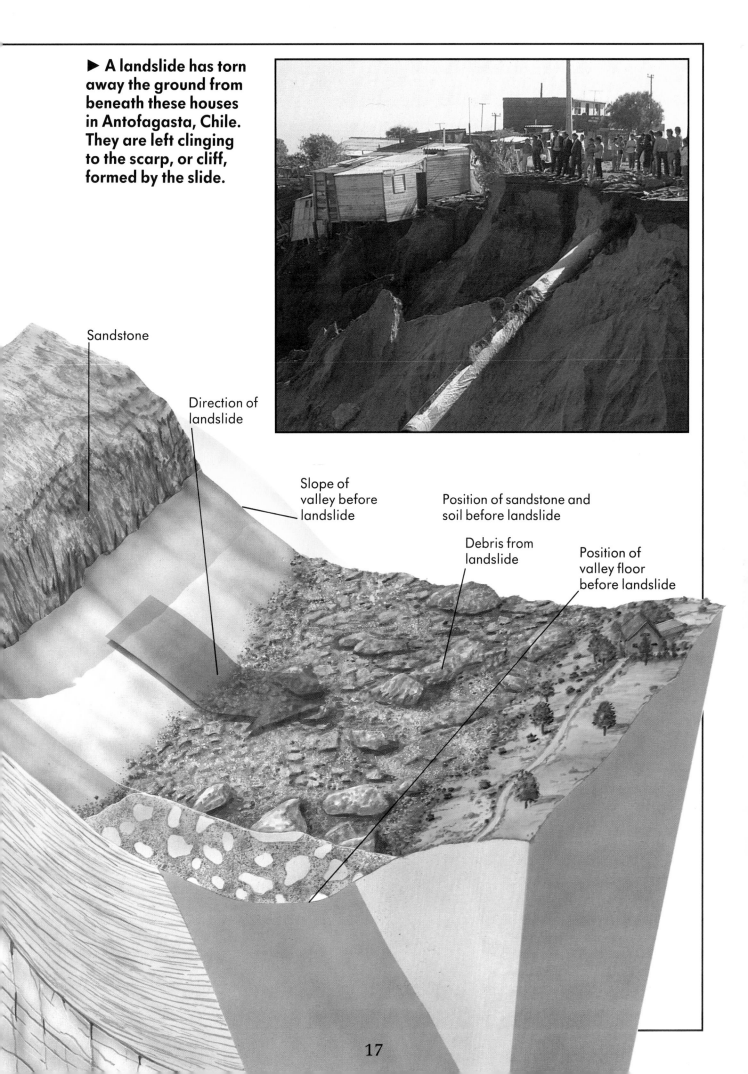

▶ A landslide has torn away the ground from beneath these houses in Antofagasta, Chile. They are left clinging to the scarp, or cliff, formed by the slide.

Sandstone

Direction of landslide

Slope of valley before landslide

Position of sandstone and soil before landslide

Debris from landslide

Position of valley floor before landslide

VOLCANIC LANDSLIDES

On May 18, 1980, Mount Saint Helens volcano in the Cascade Range of southwest Washington, erupted violently, following an earthquake beneath the volcano. As steam, hot gases, and other volcanic material shot out from the volcano, the shockwaves from the earthquake triggered off two enormous landslides on the northern side of Mount Saint Helens.

The falling rocks mixed with melting snow and ice, and the ash and dust from the volcano, forming a kind of mudflow called lahar. The name "lahar" originally referred to a volcanic mudflow in Indonesia, but it is now used to describe volcanic mudflows anywhere in the world.

Advance warnings had led to the evacuation of the area around Mount Saint Helens. But five years later, the inhabitants of Armero and other towns and villages lying in the shadow of the Nevado del Ruiz volcano in Colombia were not so fortunate. In November 1985, a violent eruption of the volcano left around 25,000 people dead and 20,000 more without shelter.

The mudflow from the volcano eventually covered an area of more than 15 square miles. It destroyed the valuable farmlands of this rice- and coffee-growing area.

▼ A lahar, or volcanic mudflow, can travel over great distances at high speed. Heat from inside the volcano helps to melt the snow and ice on its slopes. The water mixes with loose material on the volcano's slopes to form a massive tongue of mud.

▲ Several weeks before the earthquake, Mount Saint Helens showed signs of increasing volcanic activity. A large bulge which had appeared on the volcano's northern side was torn apart by the power of the eruption. Clouds of ash and dust poured into the atmosphere (above).

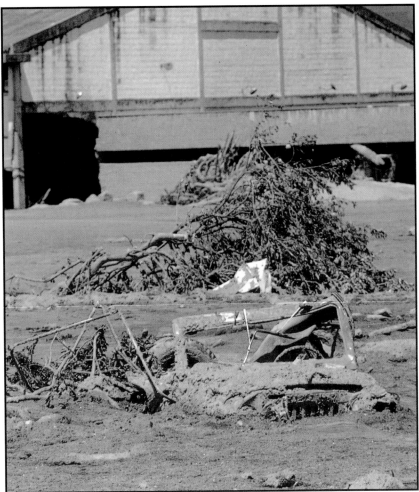

▲ Following the 1985 eruption of Nevado del Ruiz, torrents of mud raced through the streets of Armero (above) at around 100 mph. Cars were swept away, trees were uprooted, and many of the town's inhabitants were engulfed by the tide.

DISASTER STRIKES

When a landslide begins its descent, huge quantities of soil, mud, rocks, and smaller debris hurtle down a hillside. Whole villages are buried alive, and their inhabitants and livestock are swept along on a tide of mud and debris. Trees are uprooted, houses are flattened or swamped, and communications are cut off.

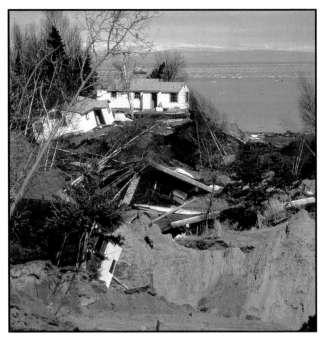

The mountainous regions of Nepal in Asia are prone to torrential rains. Here, the water-saturated soil has been eroded, exposing the layers of rock beneath. This erosion is worse in areas where large numbers of trees have been cut down. The hillsides are stripped of protection against the falling rain, and of tree roots to anchor the soil in place. Landslides glide over the smooth rock at speeds of up to 60 feet per second.

Following an earthquake in the Andes Mountains in Peru in 1970, a devastating mudflow swept through the Peruvian town of Yungay. Almost 18,000 people were buried alive as a giant wave of mud and debris about 263 feet high swamped the town.

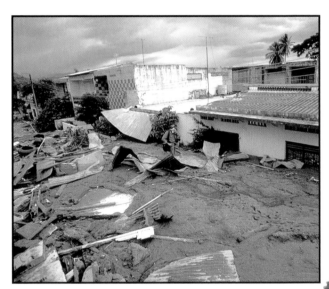

▶▲▲ Shockwaves from the 1964 Alaska earthquake caused a mudflow of clay to engulf the town.
▶▲ Mud and floodwaters wreaked havoc in the streets of Armero, Colombia, after the 1985 volcanic eruption there.
▶ The 1970 mudflow in Yungay, Peru, caused severe damage to railroad lines and roads (right).

Torrential rains

San Gabriel Mountains

Direction of mudflow

Glendora

◄ In 1969, in the area around Los Angeles, about 44 inches of rain fell in just 42 days. These rains, some of the heaviest ever recorded in the area, washed down mud and gravel from the San Gabriel Mountains. As a torrent of mud, debris, and water swept over the suburb of Glendora, over 100 people were drowned and 15,000 people had to evacuate their homes. The landslide also blocked the main highway into Los Angeles.

VAIONT, ITALY, 1963

During the night of October 9, 1963, millions of tons of rock slid down from Mount Toc in Italy into the Vaiont Reservoir below.

Three years previously, one of the world's highest arch dams was built below the mountain. Engineers working on the dam had noted that when the water level in the reservoir rose, a rocky mass on the mountainside moved downwards. As the water level increased, so did the rate at which the mass moved. Heavy summer rains swelled the reservoir, and the engineers tried to halt the rate of rock movement by lowering the water level. But the rockslide could not be stopped.

At 10:41p.m. a torrent of rocks, mud, and water raced across the river valley and up the opposite slope. As it fell back into the reservoir, a 328-foot high wave was catapulted over the 860-foot high wall of the Vaiont Dam into the valley below.

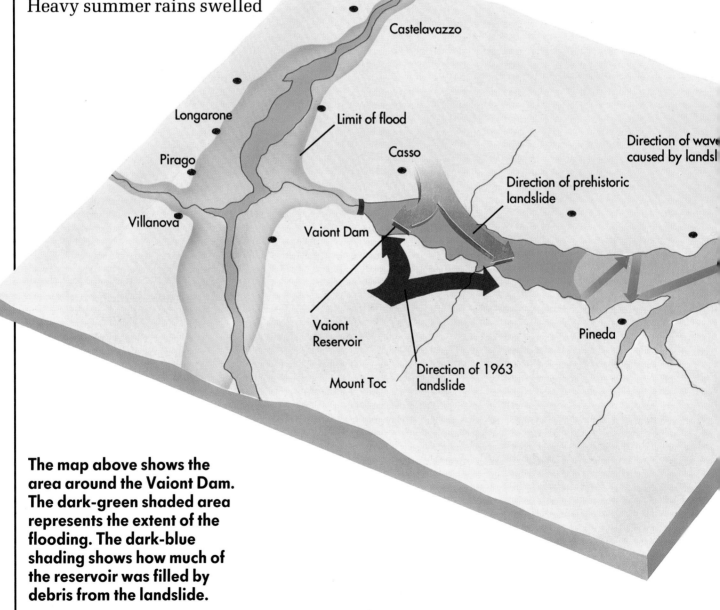

The map above shows the area around the Vaiont Dam. The dark-green shaded area represents the extent of the flooding. The dark-blue shading shows how much of the reservoir was filled by debris from the landslide.

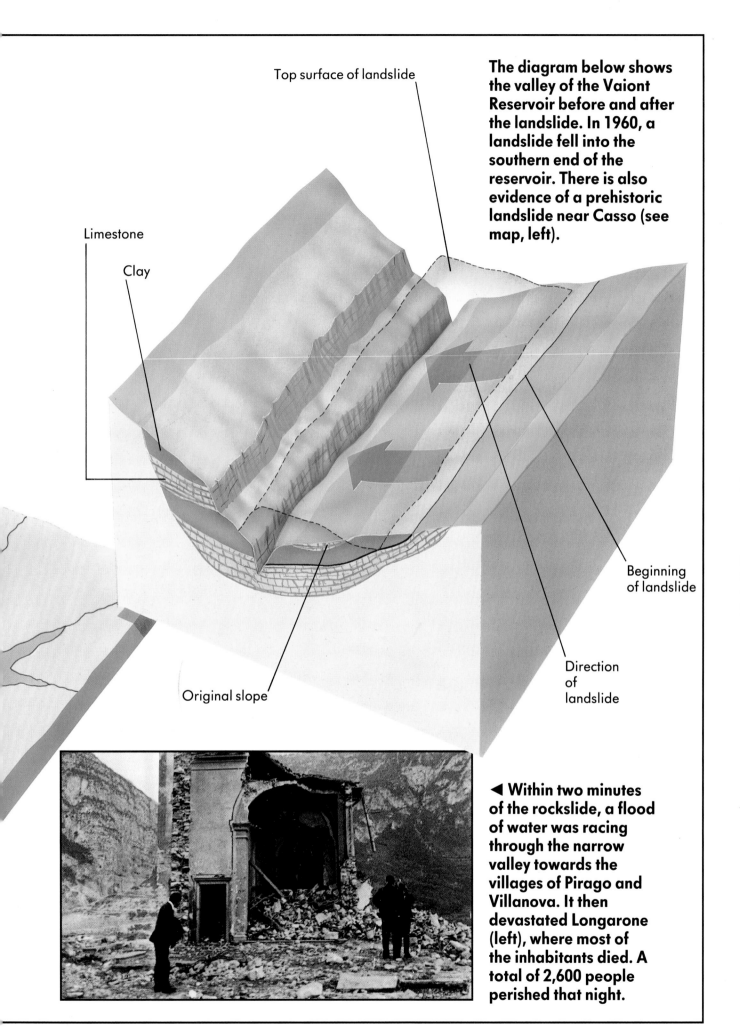

Top surface of landslide

The diagram below shows the valley of the Vaiont Reservoir before and after the landslide. In 1960, a landslide fell into the southern end of the reservoir. There is also evidence of a prehistoric landslide near Casso (see map, left).

Limestone

Clay

Beginning of landslide

Direction of landslide

Original slope

◄ **Within two minutes of the rockslide, a flood of water was racing through the narrow valley towards the villages of Pirago and Villanova. It then devastated Longarone (left), where most of the inhabitants died. A total of 2,600 people perished that night.**

23

LOOKING BACK

Evidence of avalanches and landslides dating back to prehistoric times can still be seen today. On the floor of the Saidmarreh Valley, in western Iran, lie the remains of a landslide that scientists believe occurred over 2,000 years ago. Limestone debris covers an area of around 77 square miles, and is 985 feet deep in places.

In 218 B.C., Hannibal and his army suffered huge losses from avalanches as they crossed the Alps on their way from Carthage to attack Rome. It was October, and fresh snow had fallen on a crust of old snow. Thousands of troops and animals were carried away by the avalanches triggered off as they crossed the unstable crust.

In the Middle Ages, pilgrims from northern Europe regularly crossed the Alps on their way to worship in Rome. Many believed that avalanches were acts of God, while others thought they showed the destructive power of the devil.

Landslide!
On the morning of Christmas Day, 1839, a gaping hole opened up in the top of the cliffs at Bindon in Devon, England (left). A vast chunk of rock sank down, leaving an enormous gap about 4,000 feet long and 300 feet wide. The steep slope, or scarp, down which the landslide moved, is still visible today, as are the huge lumps of displaced rock.

▶ During the winters of World War I (1914-18), an estimated 60,000 soldiers fighting in the Alps were killed by avalanches. Some of these avalanches were not entirely due to natural causes. After a heavy snowfall, Austrian and Italian troops often fired into the slopes on top of the enemy below. This action released huge dry-snow avalanches. It was the first time in military history that avalanches had been used as a weapon.

◀ Hannibal's treacherous journey across the Alps involved about 38,000 foot soldiers, 8,000 horses, and over 30 elephants. According to the Roman historian Livius, more than 18,000 men and 2,000 horses were lost during the crossing, mainly because of avalanches. The reduced numbers of Hannibal's troops contributed to their eventual defeat by the Roman army.

ARE WE CAUSING MORE?

We cannot prevent natural disasters such as avalanches and landslides. Yet we can stop or reduce the human activities that make it more likely that one of these powerful forces will strike. When large numbers of trees are cut down, to provide fuel and lumber, as well as land for farming, nothing remains to trap the rainwater and hold the surface of the slopes in place. This deforestation also removes the natural barrier to avalanches and landslides that trees and other vegetation provide.

Mountainous areas in other parts of the world are also being cleared and developed as tourist resorts. Millions of people then visit the mountains to sightsee and explore. These activities can wear away the soil and vegetation, making the area more prone to avalanches and landslides.

Road construction
Excavation work for new road building steepens slopes and removes support for the upper layers.

Quarrying and mining
As large amounts of natural resources, such as stone and coal, are removed from the ground, the surrounding land becomes unstable and may collapse.

Deforestation
The roots of trees and ground foliage hold water like a sponge, releasing it slowly into the surrounding soil in a controlled flow. When a hillside is stripped of its cover, the exposed soil erodes very quickly.

26

The threat of landslides and avalanches is increased by excavation and construction work, and by the removal of natural resources from the ground. The shape of a slope may be steepened or altered to such an extent that it becomes unstable. Heavy rains and snow add further weight to the slope, weakening the surface soil's grip on the hillside.

Reservoirs
The construction of reservoirs in narrow mountain valleys that are prone to landslides increases the risk of a flood disaster.

▲ Trees help to prevent the erosion of soil by heavy rains. Deforestation has caused this landslide in Malaysia (above). The slide has completely destroyed a main road.

Global warming
As the atmosphere traps increasing amounts of the greenhouse gases that warm the earth, changes in weather patterns may occur. Very heavy rains could increase the number of mudflows and landslides.

Forest fires

A lack of hillside vegetation is blamed for the mudflows that are common in parts of southern California. Forest fires destroy the trees and shrubs on the steep slopes around many residential areas. After heavy rains, the topsoil on the slopes becomes saturated with water and starts to slip downhill.

◄ Avalanche fences have been erected on these slopes in the Valais region of Switzerland (left). They help to break up the falling slabs of snow into smaller, less destructive chunks.

Forestation

Snow fences

Roads protected by concrete tunnels

As well as erecting physical barriers against avalanches, concrete tunnels can be built to protect roads and railroad lines. Improved warning and evacuation systems also help to reduce the loss of life from these disasters.

Trenches

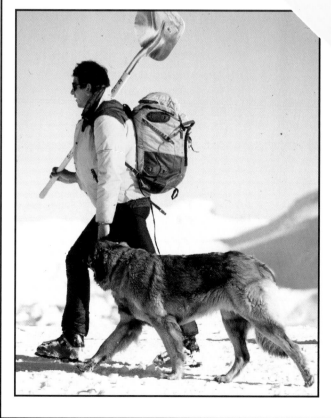

◄ Following an avalanche, trained dogs use their sense of smell to detect bodies under the snow. Long poles, called sounding rods, are pushed into the snow in the hope of touching an avalanche victim.

Trees
planted

Volcano
early
warning
system

▲ **Trees have been planted on the slopes above this mountain village to protect it against avalanches.**

Landslide prevention involves the use of trenches, pumps and wells to control water levels. New developments should be located in areas where the land is stable, or where slopes can be supported.

Controlled
quarrying

Underground drainage

Landslide barriers

WHAT CAN WE DO?

Many different measures can be taken to reduce the devastating impact of an avalanche or landslide. In Switzerland, controls include setting off explosives to release avalanches artificially, and planting triangular patches of trees above mountain villages. Antiavalanche constructions, such as trees, fences, or barriers, are used to slow down, deflect, or break up the falling snow.

Water is one of the prime causes of landslides. Measures can be taken to control and monitor the amount and

pressure of the water in unstable slopes. Underground drainage is also installed in some places to reduce the water content of landslide-prone hillsides. The lower end, or toe, of a slide can be shored up with concrete, rock or soil.

In the United States, scientists are using computers to try and forecast avalanches. The study of volcanoes can also help to predict volcanic eruptions more accurately, thus reducing the death toll from volcanic mudflows.

FACT FILE

The world's largest landslide

The largest landslide in the world in the past 2 million years is believed to have occurred in the United States. At the foot of Mount Shasta, an active volcano in the state of California, lies an area of debris covering some 175 square miles. The debris is believed to date from a rockfall that happened between 300,000 and 360,000 years ago.

Recent avalanches and landslides

1990

Peru – following heavy rains, a mudslide buried the village of San Miguel de Río Mayo in the jungle region 500 mi north of the capital, Lima. A total of 200 people were reported missing.

1991

Malawi, Southeast Africa – after flooding and a series of mudslides in the southeast of the country in March, 516 were reported dead or missing, and over 40,000 people were left homeless.

New Zealand – a huge avalanche on December 14 almost destroyed the eastern face of Mount Cook, the highest mountain in New Zealand. Thousands of tons of ice, snow, and rock hurtled down from the summit of the 12,350-foot high mountain, and smashed into the Tasman Glacier, a popular tourist spot, over 4 miles below.

1992

Argentina – torrential rain in January caused a debris avalanche of rocks, mud, and trees to destroy the town of San Carlos Minas. Sixty people were believed to have been killed.

Turkey – avalanches in February in the remote Kurdish Mountains of southeastern Turkey left more than 150 people dead. In the village of Gormec, 71 soldiers died under a single avalanche. Rescue workers were delayed by blocked roads and strong blizzards.

Israel – tons of rocks and soil fell on a cafe in East Jerusalem on February 29, leaving 23 people dead and a further 20 injured. After heavy snow and rain, part of the hillside above the café collapsed and burst through a wall onto the café below.

Avalanche control in Canada

In Canada's Selkirk Mountains, one of the world's largest anti-avalanche programs is operated along Rogers Pass, with the help of the Royal Canadian Horse Artillery. In a densely populated mountain area, the soldiers use gunfire to release avalanches and protect a 25-mile stretch of major road from 160 separate avalanche paths.

Mudflow technology

Japan has become a world leader in the study and prevention of mudflows.

Since 10 percent of the world's active volcanoes are found in Japan, many of the country's population centers are threatened by mudflows. On the side of Mount Usu on the island of Hokkaido, steel and concrete structures have been erected to trap falling rocks and slow down the flow of mud and other debris. Elsewhere, television monitors and measuring equipment help to detect volcanic activity and the onset of a mudflow or other landslide.

Volcanic mudflows

On the slopes of Mount Rainier, a volcano in Washington state, more than 55 mudflows have been recorded in the past 10,000 years.

Landslide deaths

Between 25 and 50 people die in the United States each year from landslides.

Mining waste disaster

The waste material from mining and excavating can be a serious landslide hazard. For many years, the waste from the coal mines around Aberfan in South Wales was piled dangerously high. When it became saturated with water following heavy rains in October 1966, part of the trash heap collapsed. The waste material slid downhill, killing over 144 people including 116 children in the local school.

GLOSSARY

avalanche – the downward movement of a mass of ice and snow.

creep – the slow, almost unnoticeable, movement of loose soil down the surface of a slope.

debris – loose solid material such as soil, mud, small pieces of rock, and stones.

deforestation – the large-scale removal of trees and other vegetation.

dry-snow avalanche – a huge blast of white powdery snow that travels at great speed along or above the ground.

earthquake – a violent shaking movement of the outer layer, or crust, that surrounds the earth.

earth tremor – the violent trembling of the ground that occurs during and after an earthquake.

erosion – the removal or gradual destruction of a surface by the action of water, ice, or wind. Human activities, such as tree-felling, can also lead to erosion.

eruption – the violent release of steam, rocks, dust, and ash through an opening in the earth's surface.

global warming – an increase in the earth's temperature due to a buildup of greenhouse gases, such as carbon dioxide, in the atmosphere.

greenhouse gas – one of the gases in the atmosphere that traps heat from the sun and keeps the earth warm. Greenhouse gases include carbon dioxide and water vapour.

groundwater – water that is found in rocks and soil beneath the surface.

lahar – a mixture of rocks, melting snow and ice, and the ash and dust from an erupting volcano.

landslide – the downhill movement of large amounts of rock and soil.

lubricate – to make a surface slippery.

mudflow – the rapid downhill movement of soil mixed with water.

rockfall – the downward movement, usually through the air, of rocks and other material from a cliff or steep slope.

rockslide – the slipping movement of rock along one or more surfaces.

sandstone – a coarse rock made up of grains of sand joined together.

saturated – filled with the maximum amount of a liquid or solid substance.

scarp – the steep slope left behind when a landslide falls away.

shale – a kind of soft rock made from tiny pieces of mud and clay.

shock wave – a burst of energy that is released from the center of an earthquake.

slab avalanche – the downward movement of a large chunk of solid snow.

slump – the movement down a slope of a block of rock and soil. The slumped block tilts back towards the slope.

unstable – describes a slope that is likely to start moving downward.

volcano – an opening in the earth's surface through which hot rocks and gases can escape.

wet-snow avalanche – the downhill sliding movement of large chunks of wet snow.

INDEX

Photographic credits:
Cover and pages 10, 19 top left, 20 top, and 27 top: Frank Lane Picture Agency; title page and page 20 bottom: Robert Harding Photo Library; pages 4-5 and 9: Science Photo Library; page 13 top: Paul Galvin; page 13 bottom: Charles de Vere; pages 15, 17, 19 top right, and 20 middle: Frank Spooner Pictures; pages 16 and 28 top: J. Allan Cash Photo Library; page 18-19: Natural History Photographic Agency; page 23: Topham Picture Source; pages 24 and 25 bottom: Mary Evans Picture Library; page 25 top: Roger Vlitos; pages 27 bottom and 28 bottom: Spectrum Colour Library; page 29: Geoscience Features.

PRINTED IN BELGIUM BY proost INTERNATIONAL BOOK PRODUCTION